NOBODY TIPS
A SCANDISCOPE

SOCIAL AND
ETHICAL

NOBODY

- TIPS A -

SCANDISCOPE

Discussion Guide

Includes Story, Q&A and Facilitator's Tips

ENERGYPHILE SESSION №2

ENΞRGYPHILE

Published by Energyphile Media Inc.
energyphile.org

ISBN 978-1-9991113-6-6 (paperback)
ISBN 978-1-9991113-7-3 (ebook)

Produced by Page Two
pagetwo.com

Edited by Lori Burwash
Cover and interior design by Taysia Louie
Original design concept by Christina Sweetman

Contents

PREAMBLE

WHEN I STARTED WRITING "Nobody Tips a Scandiscope," I was pretty sure the central lesson was going to be about the ethics of energy. Behind the comforts and benefits of the light switch, the gas pump or, in this case, the fireplace, there's often upstream human sacrifice.

This story's ethical message is clear. It's painfully amplified whenever I go to European cities like London. Looking at those tall chimneys, I think about the enslaved children who were forced to crawl through narrow brick flues caked in the toxic soot of 19th-century coal combustion. What agony did those children suffer? What kind of people sanctioned such a horrific practice?

But as I read more from my 1825 book *The Chimney-Sweeper's Friend, and Climbing-Boy's Album*, I realized there was a deeper question: Why did child chimney sweeping continue for a hundred years after the invention of technology that should have snuffed out the practice?

The answer lies in the selfish interests of individuals involved in the economy of chimney sweeping. Both chimney masters and the housekeepers of the homes they visited had a stake in prolonging the practice, making unfamiliar technology like the Scandiscope a tough sell.

These days, "Nobody Tips a Scandiscope" sits in the back of my mind as I think about the barriers to adopting products and processes that reduce environmental impacts of energy use. I don't look to technology as the barrier, I look to the economics of human behavior.

I hope this story resonates with you, as it did with me. And the next time you're in Europe, look up at the chimneys. You'll shake your head at the practice of using children as chimney sweeps. Shake your head too at why it took so long to save them.

Why did society **push innocent children up chimneys** for so long?

NOBODY TIPS
A SCANDISCOPE

I climb'd, and climb'd, and climb'd in vain,
No light at top appear'd;
No end to darkness, toil, and pain,
While worse and worse I fear'd.

"The Climbing-Boy's Soliloquies," *The Chimney-Sweeper's*
Friend, and Climbing-Boy's Album

THE SIMPLEST THINGS can sometimes teach us the most. That's why I want to tell you a story about chimneys.

A chimney is simple enough: it's a hollow column made of bricks, stone or cement that vents out gases from the fuels we burn. But within those flues lurk eye-opening lessons about the influences on, and implications of, our energy choices.

To get in the mood for this story, do what I'm doing as I write it: settle into a comfy chair by a crackling fire late at night. Absorb the heat and let your mind wander. Imagine you're in England's north country, where it's damp and cold. Now imagine it's January 1857.

HENRY HAWORTH was nervous.

He was standing at the gates of a 16th-century school, looking up at the roof of a large three-story stone building. The glint of the early morning moon backlit four cylindrical pots, grouped in a row on top of the chimney.

Only seven years old, Henry already knew his chimneys. He knew them inside and out. Literally. You see, Henry was a chimney sweep, although that was a benign description of what he did for a living.

In mid-19th-century Britain, most people would have referred to Henry as a climbing boy, which in my mind was a more fitting handle. That's because he didn't really sweep much. Instead, he spent his long days climbing inside fireplace flues, removing layers of soot with his head, hands and a scruffy straw brush.

Before the end of this January work shift, Henry would have to squeeze through each of those four suffocating chimneys.

Like any child, he'd learned quickly from experience: the first stack usually vented the main-floor fireplace — he dreaded that one the most. Filthy stone flues that snaked through walls for 20 meters were common. Add to that 90-degree kinks jammed with dregs still hot from the previous night's fire, and a flue seemed to go on forever, all the way up to the clay pot's narrow opening at the end of the stack.

Henry kept staring at the top of the building.

Having never gone to school, he could neither read nor write. But he could add up what scared him. Four, he thought, four of the scariest tall-pot chimneys. All to be cleaned before noon!

The boy's worrying was interrupted by his foul-tempered chimney master.

"Stop starin' and git a move on!" barked Nigel. "We got a lot to do this morning." His snarl revealed a mouth full of rotten teeth.

Henry took a deep breath of crisp winter air. It would be his last pleasure of the day. Head hanging, he limped behind Nigel through the school gates. No one cared that his left ankle was sporting a bloody scrape, inflicted by one of yesterday's merciless chimneys.

Might to Blight

Most people don't notice chimneys, but I do. As an energy tourist, I take pictures of them, especially when I travel to old European cities with rich crops of chimneys sprouting out of urban landscapes. Why? Because chimneys have been fundamental to the evolution of our energy needs.

By the 15th century, chimneys were gaining popularity in Europe as a way to separate smoke from the warmth of an indoor fireplace. Soon after, steam engines started pulling, pushing and turning, and it wasn't long before mechanization led to electric power generation. All these fuel-hungry contraptions needed a stack for blowing out the energy we *didn't* use in the process of creating economic prosperity and comfort. With the Industrial Revolution, chimneys came to signify productivity and industrial might. Now, they symbolize two of society's great energy ills: inefficiency and environmental degradation.

I admit I hadn't given chimney sweeping much thought until I bought a rare 1825 book titled *The Chimney-Sweeper's Friend, and Climbing-Boy's Album*, edited by James Montgomery. By 1825, the practice had already been in play for over 150 years, and Montgomery, a poet, newspaper publisher and tireless advocate for the helpless children, documented much of the horrific history. Intrigued and appalled by what I read, I looked for

Once you've heard the stories of the climbing children, you never look at chimneys the same way.

more materials on the climbing boys' plight. My hunt yielded "Climbing Boys," an 1857 article by an unknown author. It opens with a quote that's ominously on point: "They die in their youth, and their life is among the unclean."

Between this short essay and Montgomery's 500-page treatise, I'd added to my library two of the most depressing pieces of literature I own. Their pages are filled with heartbreaking stories of climbing boys being burned, beaten or suffocated to death. Like John Anderson. At five years old, he'd already been working for a year and a half. One day, he was sent up a flue still hot from the night before. Within five minutes, people could hear him groaning, but they couldn't reach him in time and the poor boy suffocated.

And so, I added ethics to the list of energy ills symbolized by a chimney.

Henry was the middle child of three, born to an impoverished family that desperately needed money to survive. Six months earlier, his mother had made the unimaginable decision to sell her son, a perfectly aged and physically able boy, into Nigel's apprenticeship.

"Here's three shillings fer ya," Nigel had sneered, dropping coins into her shaking hands.

"Come here, boy." The sweep grabbed Henry's arm, claiming his newly acquired asset.

"Please, Mum, please! Don't let him take me away," Henry had whimpered, straining his free arm out to his mother. But it was too late. The deal was done, and she was already retreating into the shadows of a narrow street.

With that, Nigel became Henry's chimney master or, more bluntly, his slave master.

Signing Their Lives Away with an X

Not all climbing boys were sold into slavery by their parents. Many in England were orphans offered up by workhouses, state-run dwellings for the destitute. In fact, chimney masters were often paid to take children away — for as little as $20 in today's terms — to ease the state's burden of caring for the neglected.

During the transaction, a child would have to sign an indenture, usually with an X, in front of witnesses, pledging seven years to his new vocation. In return, the boy's master was responsible for feeding, clothing and sheltering his new charge.

Many good masters fulfilled their end of the bargain, but it's well documented that plenty of unscrupulous ones fell short of delivering on their basic responsibilities. In fact, undernourished boys were desirable because they were better able to climb narrow passages. How narrow? Open a magazine. That 11-by-17-inch spread is larger than the standard flue child workers had to squeeze through.

Thinking about such conditions reminds me of *Oliver Twist*, in which Mr. Gamfield, a master sweep, discusses how stuffing burning straw up a flue motivates climbing boys: "… even if they've stuck in the chimbley, roasting their feet makes 'em struggle to hextricate theirselves." The chair-squirming truth is that Dickens' fiction is based on facts openly published at the time.

Girls weren't exempt either. Some were pressed into service if their chimney master parents didn't have sons to support the business. Their slighter frames also made them better able to fit into those tiny flues.

Nigel led Henry to the fireplace, still oblivious to the young one's limp. Before the boy got started, he took off his shoes and brought out his climbing cap. He pulled the improvised

Chimney masters were often paid to take children away — for as little as $20 in today's terms.

balaclava over his head and below his chin, to prevent ingesting the toxic residue.

Henry hated this part — preparing for the dark unknown. He wished his mother was by his side, to kiss his ankle better, to soften his fears. But she wasn't. Hoping to rid himself of memories of her, he gave his head a shake and grabbed a small straw brush. He launched himself up off the hearth's grate and pushed into the impossibly tight flue by pressing the sides with his back, elbows and knees, articulating his body like a caterpillar.

Scraping, brushing and squirming, Henry loosened soft and hard soot for the first nine meters, allowing the residue to freely fall into the fireplace. Later, Nigel would put the pitch-black soot in sacks and take it to the afternoon market, where he sold it to farmers as fertilizer.

Henry kept plodding skyward until he poked his head up to a horizontal stretch. Pulling himself around the corner on to the ledge, he found momentary relief from the force of gravity. But now he had to scrape forward, channel the soot underneath his belly, then push it back with his feet. This repetitive forward-reverse motion gradually thrust all the residue to the free-fall of the vertical segment he'd just cleared.

After a few meters, Henry's head touched a wall. Catching his breath, he sensed open space above. From experience, he knew he was about to bend upward into the final vertical section. Relief set in. Soon he would be done this first flue.

The Long Road to Saving the Children

While the energy geek in me has always been interested in European chimneys, I look at them differently now. Examining

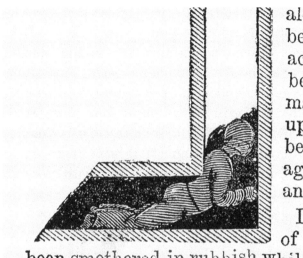

along with 1
bedded in
accumulates
becomes dil
makes a des
upper angle
becomes cor
agony of su1
and expires !
In the prese
of climbing-
been smothered in rubbish while " coring

Suffocation was the most common cause of death for climbing boys. Squeezed in a flue no larger than 11 by 17 inches, a child had no hope of escape should soot collapse, dying in "an agony of suffocation."

them inside and out, I imagine those poor climbing children and reflect on what their horrors mean to us today.

The obvious takeaway is that we must be mindful of the ethical issues associated with our energy systems. I'm not naive to what goes on behind the scenes of our Western comforts. Nor should you be. The gasoline you put in your tank may have originated from an authoritarian state that tortures its citizens. The battery in your phone, weed whacker or electric car may contain scarce metals from Africa, chipped out of a mine by a child. These are 21st-century realities.

But as I sit in front of my fireplace, I'm thinking about a bigger issue. The first ban on the use of climbing boys came in 1788. Mechanical chimney-sweeping devices were around as early as 1789. Why did society continue to push innocent children up chimneys for another hundred years?

From a callous business perspective, cost and implementation were two big reasons. Early chimney-sweeping innovations were expensive and complicated. It took until the 1870s for the cost of mechanical brushing systems to come down. Many master sweeps couldn't afford the upfront cost of a mechanical contraption. If they could, they argued it took too long to figure out how to use the device properly, especially in hard-to-sweep areas like horizontal sections.

So it was cheaper and more effective to use youngsters.

In 1803, the Society for Superseding the Necessity of Climbing Boys was founded. Dedicated to saving climbing boys from slavery, the organization also promoted the use of mechanical technologies. Encouraged by the society's formation, the Royal Society of Arts revived its languishing design competition for chimney-sweeping machines. In 1805, George Smart's "Scandiscope" won. The Scandiscope was a series of hollow rods held together by a long cord that ran down the inside. The chimney sweep connected the rods one by one, creating a long-handled straw brush he pushed up the flue, scrubbing the chimney.

Other innovations followed. In 1828 Joseph Glass introduced a similar kit that used screw-together brass rods to extend the long handle. Considered superior to the Scandiscope, Glass's tool began penetrating the sweeping market in earnest, but there were still many years to go before climbing boys were eliminated.

These innovations were paralleled by legislation that went largely ignored. Adherence to the 1788 ban was minimal, especially in rural England and Wales, where child laborers were cheapest. Authorities notionally accepted the master sweeps'

Because technology can flatten hierarchy, it isn't always desirable.

arguments against the chimney-sweeping contraptions, turning a blind eye to the legislation.

The rapidly increasing use of coal during the Industrial Revolution was another compelling reason to let the practice go unchecked. Chimneys were being built at a frenzied pace, and sweeping was an essential service for fueling prosperity.

But basic economics wasn't the primary headwind against the adoption of mechanical cleaning. As I read more about climbing boys, I realized that the persistence of entrenched interests was the greatest source of friction — the slow pace of change wasn't due to a lack of technology, it was thanks to selfish, human fixations.

Take social status, for example. In many 19th-century cultures, the privileged class didn't have much interest in changing its standing in life. Society's bottom rung, represented by individuals like climbing boys, was necessary to the preservation of class distinctions.

In thinking about this, I pause to look at the illustrations in Montgomery's book: top-hatted gentry point at and disparage

The social strata of the sweeping economy collide: the upper class looks on as a chimney master acquires a new apprentice. Resistance to change has many sources. In the case of climbing children, maintaining class distinction was one.

climbing boys. Social status is important to egos, and because technology can flatten hierarchy, it isn't always desirable.

When analyzing the potential for change, I think about the phrase "Money talks." Recognizing entrenched financial interests means knowing where the talking, or whispering, is going on.

In the business of climbing boys, there was a small underground economy of perquisites that were massive barriers to change. Housekeepers, the primary decision makers about cleaning chimneys, would receive tips and perks when their homeowner's chimney was kept in good order. If a chimney master sullied carpets and furniture with a new contraption, no tips were forthcoming. So their resistance was financially motivated, for fear their perks would be jeopardized by this disruptive technology.

Similarly, master sweeps had perks at stake. Servants and homeowners would often sympathize with pathetic-looking climbing boys, giving them tips and food at the end of a job. Both dividends inevitably ended up in the master's hands. Bottom line: nobody tipped a Scandiscope, a fact every master took into the calculus of adopting new technology — or not.

> Some wooden tubes, a brush, and rope
> Are all you need employ;
> Pray order, maids, the Scandiscope,
> And not the climbing boy.

The Every-Day Book; Or, Everlasting Calendar of Popular Amusements, Sports, Pastimes, Ceremonies, Manners, Customs, and Events, Incident to Each of the Three Hundred and Sixty-Five Days, in Past and Present Times

Henry began contorting his body upward into the next vertical section, but he still couldn't see any light above him. Relief turned to worry. No light almost always meant a blockage.

He was right.

Suddenly, a rush of soot collapsed, enveloping his head and shoulders.

Panic overtook the child. He couldn't move. Breathing was difficult. "Help!" he gasped. But it was futile — no one could hear him.

Within minutes, this seven-year-old suffocated to death, his body entombed as a wasted life, cleaning waste from an appliance that emitted wasted energy.

The last Chimney Sweeper.

A large brush made of a number of small whalebone sticks, fastened into a round ball of wood, and extending in most cases to a diameter of two feet, is thrust up the chimney by means of hollow cylinders or tubes, fitting into one another like the joints of a fishing rod, with a long cord running through them ; it is worked up and down, as each fresh joint is added, until it reaches the chimney pot ; it is then shortened joint by joint, and on each joint being removed, is in like manner worked up and down in its descent ; and thus you have your chimney swept perfectly clean by this machine, which is called a Scandiscope.

> Some wooden tubes, a brush, and rope,
> Are all you need employ;
> Pray order, maids, the Scandiscope,
> And not the climbing boy.

Copy of a printed hand-bill, distributed before May-day, 1826.

No May Day Sweeps.

CAUTION.

The inhabitants of this parish are most respectfully informed, that the UNITED SOCIETY OF MASTER CHIMNEY SWEEPERS intend giving their apprentices a dinner, at the Eyre Arms,

Although this May 1826 handbill dubbed the Scandiscope "the last Chimney Sweeper" and exhorted housekeepers to use the new tool, it was another 50 years before people ceased sending children up chimneys.

A Force to Be Reckoned With

The crackling fire beside me underscores the silence in my mind.

Think about it: it took five parliamentary edicts and a *hundred* years before the Chimney Sweepers Act of 1875 ended the era of climbing boys for good.

Looking up at the stone stack, I wonder how to end this story.

The obvious conclusion is that child labor is despicable. Or that unethical practices in the supply and use of energy are ill-advised. Both are vital lessons, but the primary business lesson lies in the troubling longevity of climbing boys: deeply rooted institutions in our society don't change as fast as technology.

A century and a half later, we're still trying to displace chimneys, stacks and exhaust pipes — the conduits of combusting fossil fuels. This despite the fact that more efficient energy processes are readily available. Climbing boys are a testament to the difficulty of uprooting entrenched interests and low-cost processes. When thinking about our energy future, across all energy systems, do we understand those ingrained interests?

Reflecting on my own experiences in the business world, I realize we talk far too much about the potential of new technologies — how new product A will clobber old-school product B — and too little about matters related to people, especially the hardiness of entrenched interests at a raw, human level.

In fact, technology and "know-how" are rarely the limiting factors in a successful business plan, or in making the world a better place. This tale of Henry Haworth and his climbing peers should remind us that self-interest can be so rigid, so strong, that it can seriously hinder the adoption of superior solutions — and abet even the worst human practices.

QUESTIONS
AND ANSWERS

Introduction

Images of child labor and appalling conditions in "Nobody Tips a Scandiscope" understandably bring out strong emotions in readers. That's fine — discomfort is a prerequisite for thinking about this very personal story.

Spend as much time as you need on the up-front questions — they'll help you deal with some of the sensitive issues. Later questions will get you drawing parallels to today's pressing issues and identifying social and ethical issues related to energy. More specifically, the questions that follow will help you:

- understand that there are behind-the-scenes consequences to your energy use

- appreciate that the day-to-day comforts you enjoy are often accompanied by ethical issues you may choose to ignore, or are not aware of

- recognize that seemingly small, personal self-interests can stack up to societal barriers that impede large-scale change for the collective good

- realize that self-interest can be so strong it can overpower government policy

As with any ethical issue, much is gray. There are few right or wrong answers — but there are some uncomfortable ones. My overarching intent with this story is to help you realize that energy transitions are mostly about social transformation. Without that understanding, it's very difficult to force changes in our energy circumstance.

Questions

1 Think about the emotions you felt while reading "Nobody Tips a Scandiscope."

   A What did you feel about the plight of climbing boys like Henry Haworth, and why?

   B Which bothered you more: that young children were forced up chimney flues or that regulations to ban the practice were ignored for the better part of a century?

2 In 2007, some multinational clothing retailers were called out for using manufacturers in India that exploited child labor. Children as young as 10 were found working in dreadful conditions for 16 hours a day. After Western media exposed these practices, the companies' response was swift: they implemented guidelines to eradicate the practice and invested in better conditions.

   A Discuss the commonalities between the plight of climbing boys and child garment workers 200 years later. What is the underlying motivation of using children?

   B Do you know where your clothes are made? Who makes them? Do you want to know?

3 The Democratic Republic of Congo (DRC) is a major supplier of cobalt, an elemental metal that is a key component for the batteries used in everything from cell phones to electric vehicles. Many human rights agencies and news outlets have exposed child labor (including children as young as six) in the DRC's artisanal cobalt mines, yet much of this disclosure goes unnoticed in our day-to-day lives. Chances are high that some of the cobalt in your battery-powered devices may have come from a mine that exploits child labor.

A Clean electricity and battery storage are seen as vital solutions to mitigating carbon emissions. Over the next two decades, the commercial pull for mined metals like cobalt is expected to rise exponentially. As this graph shows, production has already risen appreciably over the past 20 years.

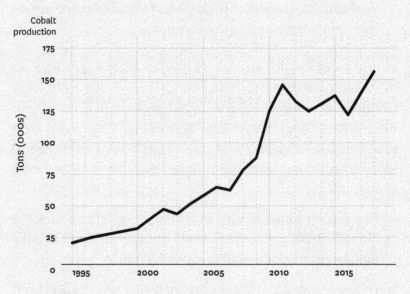

Source: *BP Statistical Review of World Energy* (see Sources Cited on page 52).

Discuss the ethical trade-offs between the demand for potentially ethically challenged metals like cobalt for electric vehicle batteries and the need to address pressing environmental issues like climate change.

B By analogy, a Scandiscope does exist for today's mining industry. Yet even though mechanized mineral extraction is technologically advanced, the use of children in mines persists. What self-interests allow this practice to continue?

C You have the choice to buy gasoline made from oil that comes from one of two suppliers: (a) a country that has strict labor laws and environmental regulations or (b) a faraway authoritarian country that abuses human rights, is highly corrupt and turns a blind eye to environmental degradation. The price to you is the same. Which do you choose?

D Would you pay more to buy your energy goods and services (for example, gasoline, natural gas, electricity) from a source you know respects human rights, labor rights, gender equality and environmental stewardship, among other ethical virtues?

4 Five main issues stalled the complete abolition of climbing boys for over a hundred years:

- the economics of self-interest
- the stratification of wealth and social status
- climbing boys were cheaper and more effective than early mechanical tools like the Scandiscope
- policies were easy to skirt
- homeowners' denial of the problem

A What are some of the forces of resistance to the transition from fossil fuels to renewable energy?

B What lessons can you glean from this story for executing an energy transition today?

5 How have government policies that target energy and the environment become muted or ineffective?

6 Denial and social resistance are often at the root of hindering change. At any time in history, the prevailing mediums of communication are persuasive factors in shaping societal attitudes. Think about access to information and opinion as you answer the following questions.

A Books that exposed the plight of the climbing boys, like Montgomery's *The Chimney-Sweeper's Friend*, did little to change attitudes. Why do you think that was?

B Why do you think the clothing retailers mentioned in question 2 acted so quickly after being exposed?

7 The story identifies wealth stratification and social status as one reason why climbing boys weren't easily replaced by a Scandiscope.

A How did wealth creation during the Industrial Revolution factor into the ongoing use of chimney sweeps? What are the implications of wealth to transitioning to a clean energy economy today?

B Developing countries, like many in Southeast Asia and Africa, accuse wealthy, already-industrialized nations, like those in Europe and North America, of being responsible for loading the atmosphere with greenhouse gases over the past 200 years. How does transnational wealth disparity affect energy transitions?

8 How have your views about energy transitions changed after reading "Nobody Tips a Scandiscope"?

9 "Nobody Tips a Scandiscope" addresses the fact that barriers to change are often hidden or unrecognized. Understanding that, what should you do if you want to push change in society, or even in your own organization?

Answers

1 A **What did you feel about the plight of climbing boys like Henry Haworth, and why?**

Many people's perceptions of chimney sweeps have probably been shaped by *Mary Poppins*. So it's no surprise that readers often tell me they were appalled to learn what went on in the 18th and 19th centuries — many had never heard of climbing boys. Everyone is deeply moved by the notion of young children being enslaved into such awful work. Emotions typically run higher among people raising young boys themselves.

1 B **Which bothered you more: that young children were forced up chimney flues or that regulations to ban the practice were ignored for the better part of a century?**

There's no right or wrong answer. It's deeply disturbing that life was considered so cheap that adults forced children up chimneys. But in countries with rule of law, it's especially troubling that government regulations were ignored, which means authorities effectively sanctioned the practice. The fact is that every complicit participant in the chimney-sweep economy — from the chimney masters who supplied the labor to the consumers who demanded the service to the government who turned a blind eye — was culpable.

💡 **FACILITATOR'S TIP**

You won't find consensus on question 1b. That's okay. The purpose is to get people thinking about the regulation of ethical issues. Encourage the group to discuss the role of the state in the ethics of energy use.

---

2 A **Discuss the commonalities between the plight of climbing boys and child garment workers 200 years later. What is the underlying motivation of using children?**

Both examples obviously involve children working in deplorable, unsafe conditions. And both involve the exploitation of child labor, historically one of the lowest-cost sources of work. Using children in factories at de minimis wage cost is effectively a form of enslavement.

Humans of any age are smart machines that can perform many repetitive, industrial tasks. If the cost of wages is low enough (or zero), there is little incentive to spend capital on machinery. These economics are especially pertinent in over-populated, impoverished countries, where parents struggling to support their families are forced to send their own children out into unimaginable working conditions.

In large part, the Scandiscope wasn't readily adopted because it was expensive to buy and maintain compared to the low-to-no cost of a young climbing boy. If a life is considered cheap — with a child laborer's life being especially disposable — and if consumers fail to hold manufacturers accountable for their practices, the decision between using humans or machines boils down to dollars and cents (or pounds and pence).

As long as there are business operators willing to discount the value of human life in the pursuit of producing the cheapest goods and services — and unwitting customers eager to acquire them — this vulnerable population will continue to be at risk.

2 B **Do you know where your clothes are made? Who makes them? Do you want to know?**

A 2017 report by the International Labour Organization estimates that 152 million children between the ages of 5 and 17 are in child labor and, of those, 73 million are in hazardous work "that directly endangers their health, safety, and moral development."

In the 21st century, such a statistic is shocking. The magnitude of child labor today suggests that we as Western consumers are quite likely enjoying the benefit of underage toil and misery.

Nevertheless, despite the pervasiveness of modern media, it may be argued that ignorance about unethical practices is come by honestly. One problem is the tsunami of information we must parse to even become aware of the issues, assuming we want to be. Amid the cacophony of issues, you have to listen hard. The chances of not hearing about questionable practices on the other side of the world are high. At a minimum, it's important to understand that our modern amenities — from clothes to energy — can be the result of uncomfortable practices hidden from our day-to-day routines.

3 A Discuss the ethical trade-offs between the demand for potentially ethically challenged metals like cobalt for electric vehicle batteries and the need to address pressing environmental issues like climate change.

Finding solutions to world-scale energy problems is never easy and is often fraught with resistance from the social force of change. The climbing boys of "Nobody Tips a Scandiscope" are admittedly an extreme example. But ignoring issues like cobalt "mining boys" may not be much different.

Does the urgency of displacing internal combustion engines with new-age batteries justify turning a blind eye to the use of child labor in foreign cobalt mines? Do the benefits of shuttering an environmentally challenging industry like coal mining outweigh the large-scale unemployment that would result? They certainly might, unless you're a coal miner.

Every decision about how we power our lives comes with trade-offs and consequences. That's because supplying and consuming energy is so foundational to society. As such, social issues are the ones that require the greatest scrutiny, the greatest understanding of gives and takes, especially when thinking about how to change up our energy circumstance.

💡 **FACILITATOR'S TIP**

Focus on the word "trade-offs." Get the group to explore the extent to which morally challenging practices justify the greater good.

---

3 B What self-interests allow the practice of cobalt-mining children to continue?

On the supply side, the short answer is poverty combined with corruption. On the demand side, it's the acquisition of consumer goods at low prices.

In places where there is extreme poverty, like the Democratic Republic of Congo (DRC), families are forced to make unthinkable choices, such as sending their children down toxic artisanal mines to dig for minerals, all for a pittance.

Coupled with that, the DRC is one of the most corrupt, unstable and impoverished countries. When there's a payoff, turning your back on ethical issues is easy to do. Unfortunately, some of the richest mineral deposits for energy storage, and most abundant reserves of oil and natural gas, lie beneath some of the world's most corrupt countries.

Far removed are manufacturers and consumers who want as-cheap-as-possible batteries to power their electric vehicles, smartphones and umpteen other electrical gadgets we take for granted every day.

Although we squirm at the notion of climbing boys in chimneys, the reality is that the primary energy resources that are vital to our comfort may still be encumbered with ethical problems we tend to ignore, and that our own self-interests are contributing to.

3 c Assuming price were the same, would you buy oil extracted from a country that has strict labor laws and environmental regulations or one that is corrupt, opaque and authoritarian?

I think it's safe to say we should aspire to transacting with trade partners that have high standards for human rights and environmental regard.

However, we know global commerce is complicated. Many industrialized nations depend heavily, if not exclusively, on oil and gas from countries with opaque practices and "business risk," which is really a polite term for corruption. So, odds are, you're using petroleum products, like gasoline, that originate from the latter kind of country.

Source: Transparency International, *B P Statistical Review of World Energy* (see Sources Cited on page 52).

Yet there's a difference between this and the climbing boy situation. Most consumers of petroleum products aren't aware of where their oil comes from. On the other hand, the use of climbing boys was visible and tacitly acknowledged, especially by society's elite, as illustrated in *The Chimney-Sweeper's Friend*.

In Dickensian England in the mid-1850s, "I didn't know that was going on" was not an excuse.

Today, distant sources of energy are hidden behind a light switch, gas pump or thermostat and totally opaque to consumers. That being so, while your answer to the choice posed in this question may be easy, acting on it may not be.

☺ FACILITATOR'S TIP

Explore whether being unaware of morally questionable practices that bring us amenities is a sufficient excuse for being indifferent about choosing a supplier.

---

3 D Would you pay more to buy your energy goods and services from an ethical source?

Notionally, the answer should be yes. And there are instances — for example, fair-trade coffee — where people pay more to guarantee such principles.

Yet, the sentiment is not universal. Money is a powerful influence, and, for many consumers, securing the cheapest product regardless of its ethical baggage is a strong force for resisting change. It's the same kind of force of self-interest described in "Nobody Tips a Scandiscope."

4 A What are some of the forces of resistance to the transition from fossil fuels to renewable energy?

This list could be long. Here are five:

- The self-preservation of incumbent fossil-fuel producers, refiners and distributors, who naturally innovate and put up barriers to preserve their market share of energy supply.

- The self-interests of downstream and peripheral businesses associated with fossil-fuel systems. For example, gas station owners, automakers, auto mechanics, vehicle financiers and many others in the transportation realm. In the electrical power world, the incumbent utilities have invested billions in coal and natural gas generating plants. Oil-burning plants still exist in some countries too. Those sunk-cost investments represent barriers to change, especially when there is scarcity of investment capital.

- Embedded attitudes about what is "normal" when it comes to established energy equipment and appliances. For example, if someone thinks it's normal and inconsequential to drive a large, inefficient combustion-engine vehicle, their propensity to change to a different paradigm — such as an electric vehicle — is likely to be weak.

- The high utility of fossil fuels, on many dimensions, at low price. Fuels like coal, oil and natural gas have high utility for the work they do. And after decades of refinement, they've become relatively inexpensive to operate. Their utility deficit is on environmental performance. When making choices, consumers weigh all dimensions of utility against price. In business parlance, the "barriers to entry" for displacing fossil fuels are high.

- Resistance — from both consumers and industry — to policies like carbon taxes. Carbon taxes are meant to raise the cost of operating fossil-fuel appliances so that their price-utility advantage is diminished relative to substitute systems powered by renewable energy. Yet any tax on the status quo is often met with social resistance, especially by those who can least afford additional stress on their household budget. The personal self-interest to resisting change is the here-and-now preservation of wealth.

Commercially, industries that rely heavily on energy as an input to operations often view carbon taxes as a competitive disadvantage. The cost disadvantage to a company arises if competitors have a lower tax, or no tax at all. Regardless, like any tax, a carbon tax can be contentious to a populace that's not willing to pay more for its energy, regardless of source. As such, opposition to fiscal levies preserves the price-utility advantage of the incumbent.

4 B What lessons can you glean from this story for executing an energy transition today?

By the 1870s, three primary forces of change had overcome resistance to ending the use of climbing boys: stricter government policy, widespread social recognition of the problem and technology that was compelling enough for chimney masters to replace their incumbent practice with it.

Therefore, the overarching lesson is you can't successfully execute an energy transition by deploying only one force of change. In this case, it was policy, social and technology forces of change that converged to bring about the transition.

When we think about shifting society away from fossil-fuel systems like oil-powered vehicles, we tend to think that policy (like carbon taxes) together with technology (like electric vehicles) are sufficient catalysts of change. They're not. As question 4a shows, there are many forces of resistance.

That's another important lesson from this story: to understand which forces of change are necessary to execute an energy transition, you first must identify all the forces of resistance, large and small.

💡 FACILITATOR'S TIP

Depending on the group, you can make question 4b more specific to their day-to-day work. For example, "When thinking about problem X in our company, what lessons can be gleaned from this story?"

---

5 How have government policies that target energy and the environment become muted or ineffective?

Your answer will depend on where you live. One commonality in energy and environmental policies is exemptions, whereby industry groups are granted leniency or outright immunity from policies. Often, industry associations argue that exemption from a levy or ban is necessary to preserve employment. For example, the coal industry has long sought exemptions from carbon taxes and pollution standards because so many people work in mines. The auto industry frequently pushes back on fuel economy targets, citing additional costs that flow through to consumers. Watered-down regulations or outright waivers can neuter a policy's intent.

Then there's the problem of changing governments. Following his 2016 election, US President Trump dismantled many Obama-era environmental policies. Provincial jurisdictions in Canada, such as Ontario and Alberta, have implemented carbon taxes under one government only to rescind them under the next. Pendulum regulations that swing back and forth depending on who's elected to office tend to hamstring energy-related policies requiring much longer than an election cycle to be effective.

Policies may also simply be ignored. This was the case with climbing boys, when policing was far less forensic in the 19th century and skirting enforcement was easy. Today, ignoring the law is more difficult in countries with rule of law and a strict regulatory regime.

But not all countries have unbiased rule of law. In countries where corruption is rife and bribery of government officials routine, it's commonplace to sidestep the law — especially environmental and labor regulations — with a wink and a nod. Corruption is an amplified form of self-interest, not dissimilar to the behavior of chimney masters, who received tips from housekeepers or expropriated them from their climbing boys.

Policies and international agreements meant to facilitate energy transitions are only as strong as the inclination of their participants to abide by them.

6 **A Books that exposed the plight of the climbing boys, like Montgomery's *The Chimney-Sweeper's Friend*, did little to change attitudes. Why do you think that was?**

A big part of the problem was that child enslavement wasn't necessarily considered an immoral practice. One could argue that because societal norms were more accommodating of such practices in the 19th century, being apathetic was easier. That's probably why society's upper echelons didn't place much value on the lives of the impoverished and underprivileged. Consequently, there wasn't widespread interest in or response to books like Montgomery's, even though he and his Society for Superseding the Necessity of Climbing Boys were very vocal in their fight for the rights of the young chimney sweeps.

Accessibility to information may have been a factor as well. Unlike the upper classes, lower classes didn't have access to books (or couldn't read at all). So while they were presumably more sympathetic to the plight of the climbing boys, being victims of the practice themselves, societal conditions were not conducive for spreading concern among the masses.

6 B Why do you think the clothing retailers mentioned in question 2 acted so quickly after being exposed?

Among the factors contributing to the retailers' swift response, social media was a big one. Clearly, corporations don't want their disturbing ethical practices exposed to the court of public opinion. Social media is now the biggest public court of all.

This scandal surfaced in 2007, when Facebook, Twitter and other social media were gaining momentum. In bringing the world closer together and bridging consumers' knowledge gap about who was making the clothes on their backs, these new communication channels helped fuel and expose the outrage.

Companies spend a lot of money building and preserving their brand value. Any threat to that is always taken seriously and dealt with expeditiously. That's why celebrity endorsers are fired so quickly if they're involved in morally contentious activity.

However, influencers are public-facing, while factory workers are behind the scenes, invisible and faceless to consumers. Workers in far-flung jurisdictions are even more removed from shopping mall patrons in the wealthy developed world. Thanks to social media, these distant child workers and their deplorable working conditions were no longer invisible.

In its mass democratization of communication, social media has not only decentralized the control corporations once held over their brand, it has also enforced greater accountability on them for their business practices.

---

7 A How did wealth creation during the Industrial Revolution factor into the ongoing use of chimney sweeps? What are the implications of wealth to transitioning to a clean energy economy today?

The middle and upper classes were the ones that owned homes and hired chimney sweeps. And throughout the 19th century, as the Industrial Revolution created economic growth and wealth, the number of chimneys, both commercial and residential, proliferated rapidly. More wealth led to more chimneys, which led to more coal (energy) use, which led to a constant demand for more climbing boys.

Transitioning away from climbing boys was a long process governed by policy, the economics of substitution and moral suasion. By the mid-19th century, an increasing number of chimneys were being swept with Scandiscope-type brushes instead of children, and by the end of the century, climbing boys were no longer.

Today, there are many parallels. Wealth and energy consumption remain highly correlated. The more money you have, the more money you spend. And the more money you spend, the more energy required to produce, deliver and use the goods and services you buy.

Higher income brackets use multiples more energy than lower — and most often with products and processes tied to entrenched energy systems like fossil fuels. For example, higher income people tend to buy bigger vehicles, homes and appliances and take vacations farther away in luxurious conditions. Giving up consumption habits is difficult, so this tendency to want more goods and experiences conflicts with the transition to greener energy. This dynamic is exacerbated in high-growth, high-population countries in Asia, where hundreds of millions of people are entering the middle and upper classes.

People naturally want more, not less. In his seminal 1943 paper "A Theory of Human Motivation," Abraham Maslow, who introduced the Hierarchy of Needs, wrote, "Man is a perpetually wanting animal." In modern society, that means people, especially those with money, are forever in need of more energy. In large part, this is why top-line energy consumption continues to rise without interruption, pulling on all systems, including fossil fuels. In the face of climate change, this increasing demand, and the choices people make to satiate it, has been elevated to a moral issue.

Wealth creation and the need for more energy are not likely to subside anytime soon. And history and human nature show that people are more inclined to stick to their tried-and-true behaviors. On top of that, non-binding political contracts like the 2015 Paris Agreement on climate change are easily ignored.

Yet the bias toward cleaner energy systems that are falling in costs has begun. We will see accelerated adoption of such systems when the Scandiscopes of tomorrow offer greater utility and become cheaper than the incumbent systems. In short, the lesson here is that, for any transition to occur, moral suasion and policy need the economics to work.

7 B How does transnational wealth disparity affect energy transitions?

Wealthy countries can better facilitate energy transitions on the supply side. Whether installing renewable power sources or a mass network of electric vehicle charging stations, it takes money to retool infrastructure toward clean energy. So, it's no surprise that wealthy countries, dominantly those in the Organisation for Economic Co-operation and Development (OECD), are leading the spending on clean energy projects. Further to the last question, "developed" countries are also the ones that have fairly flat energy growth.

Countries of lesser means, however, can't make such transitions as quickly because they don't have the money to do so or, more to the point, because their priorities are focused on raising living standards at the lowest cost.

Finally, rapidly industrializing countries like China often use the cheapest energy sources possible, which are usually also the most entrenched, notably coal for power generation.

8 **How have your views about energy transitions changed after reading "Nobody Tips a Scandiscope"?**

I began researching the history of the climbing boys several years before I penned this story. The more I collected materials and dug into the subject, the more my beliefs were amplified that energy transitions have more to do with social conditions and attitudes than new technology and government policy.

Sure, new innovative products at low cost are essential to facilitating "out with the old, in with the new." But on its own, technology is inadequate for ensuring widespread adoption of new products or processes. Same with government policy that attempts to force behavioral change.

Whenever I'm asked about the outlook for a new energy paradigm — like electric vehicles — or the demise of one — for example, the oil industry — I start by understanding potential barriers. For electric cars, concerns like "range anxiety," the fear of being stranded with an empty battery, are well documented. Then there are less obvious barriers and self-interests: Does a salesperson who's been selling combustion vehicles all their career have sufficient incentive to push a new vehicle type? Does a dealership have economic incentive to sell new vehicles that require little maintenance, when tune-ups and oil changes for gasoline-powered cars provide significant revenue? How likely is a potential buyer to consider an electric vehicle if they haven't wired their garage with a high-amperage circuit? There are many other micro-considerations not dissimilar to the reluctance of chimney masters and their customers to switch to a Scandiscope.

For me, this story more than others has changed my views about energy transitions and how to think about them. Making large-scale societal change is much harder than we think.

9 **"Nobody Tips a Scandiscope" addresses the fact that barriers to change are often hidden or unrecognized. Understanding that, what should you do if you want to push change in society, or in your own organization?**

By now, you probably understand that effecting change is much more complicated than relying on just technology and/or policy. Social, ethical, economic and other issues can loom as large barriers when trying to bring about change in the way we source, supply, use and service our energy needs. That's one of this story's enduring lessons.

Facilitating change — whether at a societal or organizational level — starts with recognizing the points of resistance within each of the six forces of change that may play a role in executing an energy transition. While the points will vary based on the circumstance, I've found some commonalities. These are good places to start.

ENVIRONMENT
Which constituency of people are affected? Is the environmental degradation being felt by enough of those people to catalyze change? Are there self-interests that will lose financially if environmental policies are made more stringent?

SOCIAL AND ETHICAL

We've addressed many social issues in these questions. Remember that social norms, self-interest and entrenched behaviors are often the greatest source of resistance, and the subtlest.

POLICY AND REGULATION

Try to identify self-interests within government that may hinder policies for change. Quite often, backroom politics stand in the way of optimal policies needed for change.

ECONOMY AND BUSINESS

Economic prosperity is a prime consideration. Anything that threatens prosperity, or is even perceived to threaten it, is a red light. At a corporate level, "prosperity" can be substituted with "profitability." Never underestimate the resistance of companies under threat of business loss. I always ask "Who is being threatened?" and "Who has the most to lose?"

GEOPOLITICS

Energy security has historically trumped all. Does the new energy system compromise consistent and reliable access to energy and all the amenities the status quo provides? If it does, and transitioning away from an energy paradigm reduces a jurisdiction's energy security, it's a non-starter.

INNOVATION AND TECHNOLOGY

As I've mentioned repeatedly, technology is rarely the limiting factor. Nevertheless, the key questions to ask are "Does the new offer better utility than the old?" and "Will anybody pay for the new if they're fully satisfied with the old?"

FACILITATOR'S GUIDE

Come Together, Move Forward

Whether it's a corporate planning session, a class discussion or a social book club, an Energyphile Session encourages critical thinking, sparks lively conversations and helps build a community equipped with tools needed to make thoughtful, well-informed decisions for a better energy future.

The questions in this discussion guide invite participants to dig deeper into the issues explored in the story and gain a broader understanding of the forces of change that affect our energy circumstance. Encouraging people to learn from the past, understand the present, then prepare for the future, these questions embody the Energyphile philosophy.

As the facilitator, you will ensure the conversation stays on track, its objectives are met and everybody walks away with a better grip on how they can move forward, while having had stimulating conversation — and some fun — along the way. This guide will help you plan and prepare for a fruitful discussion.

Planning

WHAT'S YOUR OBJECTIVE?

Start by defining your goals. Every gathering of people is different. Will this workshop be used as part of a team-building program? Is your organization facing a particular business issue you'd like your C-suite to tackle in a strategy session? Or do you just want to provoke thought at a family barbecue? Whatever the reason, bringing people together fosters a greater sense of mutual understanding.

WHO SHOULD YOU INVITE?

The occasion and objectives will determine who to include. If your guest list is less prescribed, consider inviting people from different backgrounds, companies or departments. One of the more interesting discussions we've seen included employees from across a company, both office and field staff. Another robust session saw a group of accomplished friends gather around a dinner table.

HOW MANY PEOPLE?

Aim for 6 to 10 people, preferably not more than a dozen. Any more and you can separate into breakout groups. Smaller groups allow for more questions to be covered with greater depth.

HOW LONG?

A good length is 2 to 3 hours. If you want to offer a longer workshop, consider breaking it up into discrete sections — for instance, cover all policy-oriented questions in one — or incorporating other activities, like having groups research different questions and report back.

WHERE?

Of course, there's always the boardroom, but think about leaving the office. Getting people out of their usual environment and routine turns it into a more social outing and may help shift the dynamics. Many coffee shops and restaurants have private rooms. The library probably rents rooms for free, or there may be innovation hubs that offer space. Check out community centers or coworking spaces for rentals, too.

If you're venturing outside the office, ensure the space can accommodate your tech requirements and other amenities, such as catering. If you're rounding up a geographically scattered group, try Skype, Zoom or another video-conferencing app.

Preparing

Use this checklist of tasks and suggested timeline to prepare for the day of discussion.

TASK	LEAD TIME	COMPLETED
Book the venue and any required tech.	2 months	
Invite participants (request RSVP).	6 weeks	
Order hard copies of the discussion guide for participants from Amazon or Indigo.* (If participants are responsible for purchasing their own, ensure they've ordered *at least* two weeks in advance.)	6 weeks to 1 month	
If you plan to record the workshop, take notes or generate action items, delegate someone to be responsible for that.	1 month	
Order catering.	3 weeks	
Distribute discussion guide to participants.	2 weeks	
Email the energyphile.org link to the story so people can read its vignettes and listen to the audio version. (Some will prefer audio over print.)	2 weeks	

* For orders of 10+ copies, Energyphile offers a discount. Your order must be received at least six weeks before your event. Contact hello@energyphile.org.

TASK	LEAD TIME	COMPLETED
Email any other supplementary material participants should read in advance (news stories, internal documents).	2 weeks	
Read the introduction to the questions to understand their general considerations and themes.	1 week	
Read the questions, prioritize them based on your objectives and allotted time. Add any of your own.	1 week	
Organize the tech and tools you'll need (whiteboard, flip chart, markers, sticky notes, laptop, projector, screen).	5 days	
Finalize other material you intend to use (PowerPoint deck, handouts).	4 days	
Send a reminder to participants to read the story and questions (but not the answers!). If you plan to cover select questions, you may want to let participants know which to focus on. Remind them to bring paper/pen or laptop for note-taking if they desire, and of day, time and venue. If you plan to record the session, note that, too, in case anybody has concerns.	3 days	
Refresh yourself with the questions you've identified as your priorities and determine how long to spend on each. Think about how you'd like the discussion to unfold.	1 day	
Make name tags or cards for all participants if they don't know one another.	1 day	

During

Kick things off by stating what you want participants to gain from the discussion: Learn takeaways they can apply to their day-to-day role? Determine a course of action for a specific strategic initiative? Or is it purely meant to get brains working and people talking? Ask the group what they hope to get out of it, too.

Cover the housekeeping considerations: how long the discussion will last, format, when breaks will happen, where washrooms are. If you intend to record, remind people of that.

Ask everybody to briefly introduce themselves: name, where they work/what they do, what they hope to learn.

Establish the ground rules:

- Respect all points of view — maintain an open, supportive forum for every person to express their thoughts and to learn from one another.

- One speaker at a time. Give people space to speak.

- Avoid side conversations or other interruptions.

- There is no right or wrong. Disagreement is welcome but don't make it personal. (Address the *idea*, not the person who shared it.)

- Mute your phones.

As you guide the group, weave the past, present and future throughout the conversation. Remember the Energyphile philosophy: learn from the past, understand the present, prepare for the future.

If action items will result from the discussion, ensure these are documented as you go.

As the discussion progresses, watch the time to ensure you're sticking to the pace you've determined.

Allow about 15 minutes at the end to review the action plan and summarize the discussion's main takeaways with the group. What were their most surprising "Ahas"? Did their views change as a result of the discussion?

TIPS FOR MODERATING

Your role is to keep the discussion on track while creating an environment that ensures all participants feel comfortable to speak their thoughts, even if dissenting. These tips can help you do that.

People may need time to warm up. If that first question has you facing a silent room, try paraphrasing it or suggesting other ways they may come at it. If you know someone has expertise or interest in the area, invite them to share their thoughts if they're comfortable. You could also move on to another question — perhaps one that relies more on subjective takes than deep subject knowledge — and return to the first later.

Guide the conversation to keep participants engaged and focused (very important!). People can easily get sidetracked when exploring contentious questions — make sure they stay on topic.

If the discussion does become difficult or tense, pull it back to the story to re-establish common ground and remind people of the lesson.

Direct the conversation, but don't dominate it. Stay neutral and focused on listening, rather than offering your own opinion.

Make sure everyone has a chance to be heard — and understands the gift of listening. This may mean you have to clear space for people. Moderate frequent contributors if they start to dominate, and stay attuned to the quiet participants. Watch body language. Does somebody seem uncomfortable with a topic or speaker? Does a normally silent person look like they have something to say? Use those cues to steer the conversation.

Encourage people to unpack their responses when appropriate. Prompt them with questions like "How?" "What leads you to that view?" "Can you give me an example?"

Use natural segues or lulls in conversation to move on to the next question. If the discussion shows no signs of winding down, blame the clock for cutting it off and proceeding to the next question.

After

Within a week, follow up with notes, the recording, action items and anything else promised.

If you're considering offering such sessions regularly, send participants a survey to gauge what worked, what didn't, areas for improvement or change and preferred next topics or stories.

Finally

Enjoy the discussion!

SOURCES CITED
AND
IMAGE CREDITS

Sources Cited

Page 20: Cobalt Production
 BP, *BP Statistical Review of World Energy 2019*
 j.mp/bp2019review

Page 25: "A 2017 report by the International Labour
 Organization estimates ..."
 International Labour Office, *Global Estimates of Child
 Labour: Results and Trends, 2012–2016*
 j.mp/ilo2017

Page 28: Oil Production and Business Risk
 Business risk map from Transparency International,
 Corruption Perceptions Index 2018, licensed under Creative
 Commons BY-ND 4.0
 j.mp/tirisk2018

 Production stats from BP, *BP Statistical Review of World
 Energy 2017*
 bit.ly/bp_energy

Image Credits

All objects shown are in the collection of Peter Tertzakian.
Photography by Peter Tertzakian: page 4. Other images scanned
from original sources.

ABOUT THE AUTHOR

THE QUINTESSENTIAL ENERGYPHILE, Peter Tertzakian has devoted his career to energy, first as a geophysicist, then as an economist and investment executive. He's written two bestsellers — *A Thousand Barrels a Second* and *The End of Energy Obesity* — and is sought around the world as a trusted, engaging speaker. Energyphile is the culmination of his passion and knowledge.

DISCOVER MORE AT
ENERGYPHILE.ORG

- Let your curiosity wander in the PhileSpace museum
- Explore artifacts in the Energyphile collection
- Hear the stories come to life in the audio productions
- Check out new stories and content

Continue the conversation with Energyphile Sessions

If you enjoyed this book, explore others in the Energyphile Sessions series. Covering a range of issues, these discussion guides will get you thinking about the business of energy in a whole new way. More are always being added, so check back regularly.

energyphile.org/sessions